A WINTER'S WRATH

THE ICE STORM OF 2009

BY JIM McCARTY

THE
DONNING COMPANY
PUBLISHERS

DEDICATION

This book is dedicated to the indomitable spirit that inspired those early pioneers to form electric cooperatives in the first place, and that lives on in all of those men and women who rise to the occasion and restore power under the most difficult circumstances. When Mother Nature sends her worst, we know you will always send your best.

Front cover photo: Barton County Electric Cooperative linemen David Engelage and Rodney Onstott survey a heavily damaged line owned by Ozark Border Electric Cooperative. The two were part of the largest mobilization of help ever sent to a disaster area in the seventy-five-year history of rural electrification in Missouri. (Photo by Jim McCarty)

Back cover photo: Jeff Wasson braves the elements at the top of an icy pole. (Photo by Diana Bradford, White River Valley Electric Cooperative)

The Donning Company Publishers
184 Business Park Drive, Suite 206
Virginia Beach, VA 23462

Steve Mull, General Manager
Barbara Buchanan, Office Manager
Wendy Nelson, Editor
Amanda D. Guilmain, Graphic Designer
Derek Eley, Imaging Artist
Tonya Hannink, Marketing Specialist
Pamela Engelhard, Marketing Coordinator

Steve Mull, Project Director

Cataloging-in-Publication Data
McCarty, Jim, 1962-
A winter's wrath : the ice storm of 2009 / by Jim McCarty.
p. cm.
ISBN 978-1-57864-568-8 (hard cover : alk. paper)
1. Ice storms--Missouri. 2. Electric power systems--Natural disaster effects--Missouri. 3. Emergency management--Missouri. 4. Disaster relief--Missouri. I. Title.
QC926.44.M8M33 2009
363.34'926097789090511--dc22
2009015417

Printed in the United States of America at Walsworth Publishing Company

Linemen work to repair Ozark Border lines damaged by heavy ice. (Photo by Robert Henson, Ozark Border Electric Cooperative)

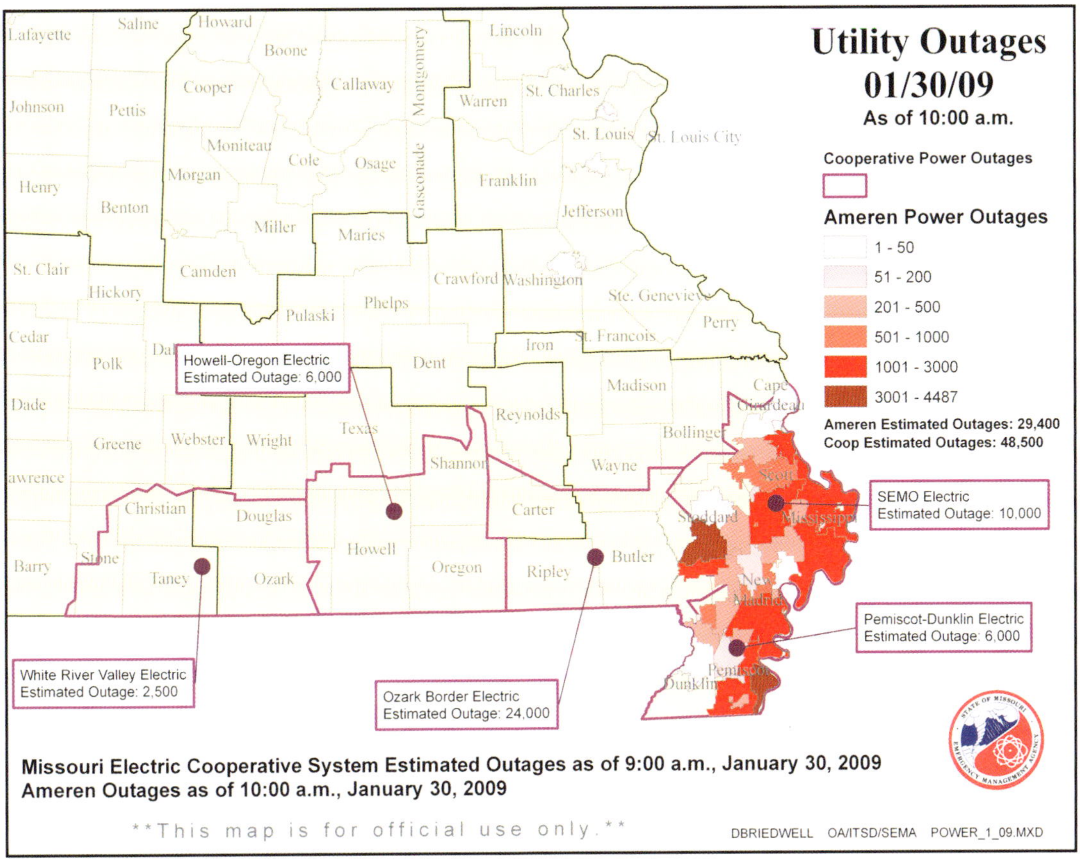

This State Emergency Management Agency situation map from January 30, 2009, shows outages for the different utilities at that time.

TABLE OF CONTENTS

FOREWORD

Governor Jay Nixon

Two weeks after I was sworn in as governor, our state faced a natural disaster that crippled a large area of southern Missouri, caused more than $193 million in damage, led to eight deaths, and, at its height, left more than 135,000 homes, businesses and other buildings without power.

But the response to the ice storm of January 26–28 also demonstrated how Missouri and its people pull together in a time of need. Even as the storm approached, preparations began for the response.

I knew that local officials in the area being hit would quickly need resources from the state to help their citizens, so I activated the Missouri State Emergency Operations Plan the first night of the storm so that various state agencies could coordinate that help.

Then even as the final wave of the storm passed, I mobilized the citizen-soldiers and airmen of the Missouri National Guard to provide assistance.

When I arrived in Sikeston the day after the storm with General Stephen Danner, the new adjutant general of the Missouri National Guard, the extent of destruction and the scope of the task at hand were evident. Everywhere we looked, we saw utility poles snapped off and tree limbs down on houses and streets.

As bad as the situation was in Sikeston, we knew that it was equally bad in other parts of Missouri.

But we also saw on this trip and on other trips to Poplar Bluff and the Bootheel how the preparation and hard work of the Missouri National Guard, the State Emergency Management

Agency, rural electric cooperatives, municipal utilities, investor-owned utilities, local officials, emergency responders, and many others combined to rebuild the power grid and restore normal activities as quickly as possible.

Missouri Governor Jay Nixon visits with a Red Cross volunteer at a shelter in Kennett. (Photo courtesy of Missouri Governor's Office)

I knew that even as thousands of Missourians took refuge at emergency shelters across the area, thousands of others were working sixteen-hour days to restore power, clear roads, and check door-to-door on people's well-being. Local utility crews were joined by crews from around Missouri and from states as far away as Minnesota to repair downed lines, many times under dangerous working conditions.

I saw firsthand during a visit to a church that was making meals how a volunteer network of thousands of other Missourians was giving its time and resources to provide both for emergency workers and for those in need.

At times of disasters like these, we also need resources from the federal government to recover. The president granted my request to have twenty Missouri counties declared as major disaster areas, meaning that now the state of Missouri, counties, and local agencies are entitled to seek millions in federal funds for disaster-related costs.

The uncertainties of Missouri's weather guarantee that from time to time, we face disasters such as these.

But what I saw in the aftermath of this storm reaffirms my confidence—both in the ability of our state to provide swift and effective assistance to those who are on the scene, and in the tireless efforts of Missourians who will step forward during times of crisis to help their fellow citizens.

—Missouri Governor Jay Nixon

ACKNOWLEDGMENTS

In writing a book about the monumental efforts of thousands of people, it is inevitable we will forget to thank someone who helped to bring this project to fruition. Nevertheless, and at the risk of leaving someone out, I would like to thank the following people who helped make this book possible:

Scott Holste, Governor Nixon's office
Susie Stoner, State Emergency Management Agency
Christi Miller and Jim Robertson, Boone Electric Cooperative
Amy Rosier and Dianna Bradford, White River Valley Electric Cooperative
Myles Smith and John Thomason, Howell-Oregon Electric Cooperative
Josh King, New-Mac Electric Cooperative
Glen Cantrell, SEMO Electric Cooperative
Pat Brown, Pemiscot-Dunklin Electric Cooperative
Teresa Fansler and Robert Henson, Ozark Border Electric Cooperative
Nancy Southworth, Linda Putman, and Kevin Hopper, Associated Electric Cooperative
Danny Campbell, Allgeier Martin
Amy Branson and Stephanie Gaffke, Association of Missouri Electric Cooperatives

More than seventeen thousand poles like this one were destroyed by the heavy weight of ice. (Photo by Jim McCarty)

CHAPTER 1

TRIAL BY ICE

Ice storms are nothing new to Missouri's electric cooperatives. In January 1949, freezing rain fell for three days across Missouri. In those early days, members sent letters or postcards to the co-op offices to let them know that power was out. One southwest Missouri man, perhaps less patient than his neighbors, rode his horse twelve miles to tell the co-op a tree had fallen through lines near his home.

A Callaway Electric lineman fights his way through fallen limbs during the ice storm of 1949. (Photo courtesy of Callaway Electric Cooperative)

The 1949 storm marked the first time the state's electric cooperatives put into action their newly developed mutual aid plan, which was adopted by the Association of Missouri Electric Cooperatives at a meeting held in December 1948. Under the plan, any system could ask for help restoring power from electric cooperatives that were out of harm's way.

Ironically, Boone Electric Cooperative Manager R. J. Martin, who introduced the plan at the statewide meeting, was the first to use it. His first call went to Farmers' Electric in Chillicothe, which sent one truck and three men. Later, realizing the damage was more severe, Martin asked for more assistance. Because phone lines were also down, his call for help was relayed by the Missouri Highway Patrol radios to Intercounty, Black River, and Scott-New Madrid-Mississippi (now SEMO) electric cooperatives. Later Howell-Oregon and Three Rivers electric cooperatives sent crews to assist at Callaway Electric.

A lineman works to repair damage near Fulton caused by the ice storm of 1949. (Photo courtesy of the Kingdom of Callaway Historical Society)

In fact, on several occasions, Missouri co-op line workers have headed to Alabama, Louisiana, and Mississippi to repair hurricane damage and as far north as South Dakota following an ice storm. When needed, other states have responded by sending help to Missouri.

In 1984, nearly 250,000 electric cooperative members were left without power in the wake of a February ice storm that primarily affected systems south of Interstate 70. A month later, north Missouri got its turn when ice fell north of the interstate line. That one knocked out power to more than 100,000 homes.

Ice storms have taken place as early as October and as late as March. They have affected few people and as many as 500,000. That was the case in November 2006 when half a million St. Louis

Linemen work the old-fashioned way to repair lines for Three Rivers Electric during the ice storm of 1984. (Photo by George Laur)

residents were without power. With their service areas largely spared from outages, electric co-ops sent fifty-three linemen from fourteen systems to the repair effort. This marked the second time electric cooperative linemen worked side by side with their counterparts at Ameren UE, an investor-owned utility based in St. Louis. The first time came four months earlier following a wind storm that struck St. Louis.

In more recent memory, the ice storm of 2007 will long be remembered by those who suffered through it. That storm cut an identical swath across Missouri as the 1949 ice storm, entering in New-Mac Electric's service area in the southwest corner and exiting just north of St. Louis where Cuivre River Electric Cooperative serves.

Comparisons between that storm and the one that hit in 2009 are telling: In 2007, 120,000 co-op members lost power. Damage was estimated at $52 million and the recovery effort prompted what was then the largest mobilization of manpower in the state's rural electric history, with thirty Missouri systems and fifty other co-ops sending 2,600 line workers to help. Repairs took fourteen days.

Fewer members, 64,000, lost power in 2009 than in 2007. And the storm affected fewer distribution cooperatives—five versus fifteen in 2007. But in terms of sheer damage, 2009 clearly takes the cake. Individual systems suffered more than the $52 million overall total in 2007. The recovery effort for the 2009 storm will probably top $200 million when all the bills come in. More line workers, 3,337, were sent into the fray than ever before. And the 17,000 poles that were replaced far exceeded the number from any previous storm.

No matter what caused the damage, or how bad it is, one common denominator can be found: electric co-op employees will rise to the occasion and do whatever it takes to get power restored. And those co-ops with outages will never be alone.

Farmers' Electric linemen Jack Boley and Don Smith work to restore service following a north Missouri ice storm in 1994. (Photo courtesy of Farmers' Electric Cooperative)

New-Mac Electric Cooperative Foreman Robert Nunley works to remove ice from one of many lines damaged by the ice storm of 2007. (Photo by Jim McCarty)

CHAPTER 2
AT THE CENTER OF MIRACLES

Shock and awe is a military doctrine based on the use of overwhelming power. Its intention is to create a feeling of helplessness in the enemy and destroy their will to fight back. Employees of Missouri's electric cooperatives who found themselves in the path of a massive ice storm that struck on January 27 and 28, 2009, know all about shock and awe.

That also was the experience for 64,000 rural Missourians who suffered through one of the most damaging acts of nature ever to hit the state in the waning days of January 2009. Some members would spend three weeks without electricity before the final outages were repaired. And mop-up work would go on well into the summer months.

On January 26, a Canadian cold front pushed into Missouri. When it collided with moisture from the Gulf of Mexico in the southern part of the state, the stage was set for a disaster of epic proportions. The storm dumped several inches of freezing rain on some parts of the Ozarks, while farther north up to four inches of sleet and seven inches of snow made travel treacherous if not impossible. By January 30, six storm-related fatalities had already been reported in Missouri—that number would climb to eight before the disaster was over.

This hand lends scale to the size of the ice that accumulated on power lines in southeast Missouri. Wind created the strange texture on the ice. (Photo by Jim McCarty)

"It was just an animal," Tim Davis, operations manager for Pemiscot-Dunklin Electric Cooperative, said of the ice storm. "People could look around and see it wasn't going to be a quick fix. People are still amazed at what happened here. They look around and can't believe it."

Once electric cooperative employees came to grips with the shock of hearing thousands of poles and trees crash to the ground, and the awe at the destruction nature had wrought, the task of putting things back together began in earnest.

From Branson to the Bootheel, five distribution cooperatives saw significant destruction with as much as four inches of ice forming on lines. The heavy weight of ice, estimated at more than 10,000 pounds per span, snapped poles in half, caused wire to sag, and toppled trees into power lines.

Broken wires dangle from one of the few crossarms that came through the storm intact. (Photo by Jim McCarty)

White River Valley, Branson; Howell-Oregon, West Plains; Ozark Border, Poplar Bluff; SEMO, Sikeston; and Pemiscot-Dunklin, Hayti all saw significant outages. Worse, this storm brought down large portions of the transmission grid owned by M&A Electric Power Cooperative.

In the flat Bootheel, power lines could be seen toppled in a row all the way to the horizon. Shattered poles, twisted steel hardware, and tangled conductor was all that remained of thousands of miles of power lines that normally kept the lights on for cooperative members in the southern part of the state.

The storm also knocked off line three power plants owned by Associated Electric Cooperative. That the electric cooperative grid took this huge hit to its generation and transmission capabilities, yet continued to function elsewhere, is a great testament to the men and women who designed it.

Across the state, the scale of destruction was met by an equally epic effort to restore power as quickly as possible.

"This was our Katrina," Rob Land told the board of directors at the Association of Missouri Electric Cooperatives, referring to one of the most destructive hurricanes ever to hit the United States. Missouri's electric cooperatives were among the first electric utilities to respond to a call for help in the wake of the hurricane. Land, who coordinates finding help for electric cooperatives in harm's way, was practically on his hands and knees as the board met, begging for another eight bucket trucks and crews to meet a request from hard-hit SEMO Electric.

Lines of trucks could be seen on just about every road in the ice storm-damaged area. Every co-op in Missouri that was not affected by the storm sent help, as did co-ops in Illinois, Iowa, Mississippi, and Louisiana. (Photo by Jim McCarty)

Land and his staff convinced cooperatives across Missouri and from Louisiana, Mississippi, Illinois, and Iowa to send crews to the Bootheel and the south-central Ozarks. When that wasn't enough, they worked with St. Louis-based Ameren UE to retain contract linemen released by the investor-owned utility. Ameren CEO Tom Voss committed the full resources of the utility to assist the cooperatives, offering materials and logistical support if needed.

U.S. Representative Jo Ann Emerson, whose district includes most of the affected area, and Senator Kit Bond personally met with managers of the electric cooperatives to ensure they had all the federal help that was needed. Governor Jay Nixon, whose term in office had just officially begun, took part in State Emergency Management Agency conference calls and traveled to the storm center to see the relief efforts firsthand.

In all, more than 3,337 people worked to restore power at the six systems involved in the ice storm. Pemiscot-Dunklin Electric Cooperative, which serves the two Bootheel counties, alone received more than a thousand additional helpers.

A house in Butler County served by Ozark Border Electric Cooperative waits for service to be restored in the early days of the ice storm. (Photo by Jim McCarty)

Besides the typical repair crews sent to speed restoration, Pemiscot-Dunklin also benefited from the help of a veteran operations manager on loan from Crawford Electric Cooperative, Jerry Wellington. Other cooperatives sent operations managers and area supervisors to coordinate the efforts of the thousands of additional line workers. Warehousemen on loan from other cooperatives helped ensure the crews had the materials they needed. Employees from the firms that deliver materials to electric cooperatives

One of many electric cooperative poles shrouded in ice. (Photo by Jim McCarty)

set up shop in the cooperative offices, speeding the delivery of truckload after truckload of supplies that came in around the clock.

Those who observed the recovery effort described the work as nothing short of miraculous. Perhaps Pemiscot-Dunklin's Davis, who spent three weeks at ground zero for the storm, said it best: "I believe in my heart that we were right at the center of a miracle. God had his hand on these people. It was miraculous what happened. There were a lot of people here who acted real professional."

Added Barry Hart, CEO of the Association of Missouri Electric Cooperatives: "This was the worst disaster to ever hit the state. It's hard to imagine what these co-ops went through. But we are very proud of what our co-ops did to restore power under the worst possible conditions. It will always be remembered as a great story of hard work and dedication."

Ozark Border Electric Cooperative crews worked late into the night to get power back on as quickly as possible. (Photo by Jim McCarty)

In some areas, trees laden with ice were a major headache. (Photo by Jim McCarty)

Pemiscot-Dunklin's Tracy Smith, right, thanks Citizens Electric Manager Tony Campbell for the supplies he brought to the cooperative during the ice storm. Besides a truckload of snacks, Citizens also sent crews and equipment. (Photo by Jim McCarty)

A stop sign framed by downed wires seems to echo the feelings of Missouri residents hit by the ice storm. (Photo by Jim McCarty)

CHAPTER 3

ASSOCIATED ELECTRIC COOPERATIVE, SPRINGFIELD

Normally, ice storms do not affect Associated Electric Cooperative, the generation co-op owned by electric cooperatives in Missouri and parts of Iowa and Oklahoma. Associated is tasked with operating the power plants that provide electricity to cooperative members. It dispatches hydropower from federally owned dams in Missouri and Arkansas and buys the entire output of Missouri's first three wind farms.

"Usually the transmission system stays up," said Duane Highley, director of power production at Associated. "And we are completely unaffected. That's always the surprise from people when we see them and there's been a major ice storm. They will say, 'I guess you have really been working long and hard.' I feel sheepish because we haven't had to do anything. That's normally what happens. All those previous ice storms it was pretty much life as usual for us."

A helicopter was used to reach the highest point on the damaged 500 kv line owned by Associated. (Photo by Danny Campbell, Allgeier Martin)

This time was different. As the transmission lines came crashing down, power being generated at Associated's New Madrid Power Plant, one of the largest plants in the state, had nowhere to go. Likewise for the gas-fired St. Francis Power Plant located at Glennonville in southeast Missouri. Power from Associated's Dell Power Plant in Arkansas could not be moved into Missouri.

"In the past we've seen some terrible devastation to the distribution systems, but on our system we've never seen this level of damage," said Associated CEO Jim Jura. "The most impressive thing I saw was how the six G&Ts worked together to get these critical lines back in service. This was quite an event for our high voltage system."

While Associated staff prepares for all kind of emergencies, the massive scale of this storm revealed problems that no emergency plan could have imagined. In short order, all three plants were isolated from the grid. "We knew we could expect distribution lines to go down in this kind of ice storm, but it's pretty much unprecedented to have this horrible devastation to transmission lines," Highley said.

Power plants go off line all the time. Normally they come down for maintenance and other planned outages. Occasionally some piece of equipment will fail. Nearly always, the interruptions are brief and the consumer never knows they happened.

One of many new poles bearing the AECI of Associated Electric Cooperative waits to be hoisted into place. (Photo by Kevin Hopper, Associated Electric Cooperative)

With three plants off line, Associated staff quickly saw an unprecedented crisis looming. The Associated transmission grid, which serves Missouri and parts of Iowa and Oklahoma, has often been called the envy of the nation. With many interconnections, multiple power plants using a diverse portfolio of fuel sources, and many miles of redundant transmission line, the grid has been tested in many ways and always held up.

Even with New Madrid's 1,200 megawatts of power unavailable, and the smaller Dell and St. Francis plants off line, power still flowed to members unaffected by the storm. The concern was with getting these plants back on line.

In southeast Missouri, the St. Francis plant is considered the blackstart plant for New Madrid. In the event New Madrid has to shut down completely, it takes electricity to get it restarted. When St. Francis was built, designers included two small diesel-powered generators that could produce the power needed to get one of the plant's two generation units restarted. From that unit, the second could be restarted, and that power would be sufficient to get New Madrid back on line.

"We have emergency diesel at St. Francis but we only have enough diesel to keep those running for a few hours," Highley said. "The thinking was, we don't need more than eight hours of run time on a typical system blackout. We never thought of a case where you have a blackout of the transmission system simultaneously with the inability to deliver fuel."

With the emergency diesel generator drinking fuel at an alarming rate, Associated staff made resupplying St. Francis with fuel their top priority. "We couldn't get trucks on the road. We couldn't get people into the plant to fill the trucks because they couldn't get on the road. They were too icy," Highley said of the situation. Ironically, one of the hazards keeping trucks off the road was a plethora of downed power lines blocking roads in the early hours of the storm.

Associated put in a call to the Association of Missouri Electric Cooperatives. The call came in during a conference call briefing at the State Emergency Management Agency. AMEC's Rob Land cut short his report to the agency and got its acting director on another line. Instantly recognizing the seriousness of the situation, the state put its resources at Land's disposal. The National Guard and MoDOT got to work clearing roads, with Adjutant General Steve Danner himself helping with the situation.

A farmer near the plant brought enough fuel to extend the run time for a few more hours. MFA Oil came to the rescue, but there was another problem. There was no electricity to power pumps to fill the truck.

This tower supporting Associated's 500 kv line suffered a broken arm from the heavy weight of ice. This line was the first 500 kv line ever built by an electric cooperative. (Photo by Kevin Hopper, Associated Electric Cooperative)

"We were in a race against time there at St. Francis," Highley recalls. "If the backup diesel generators came completely off line, then we have no way to keep heat to all the piping on the boilers. So we were in a very real crisis there believing that if we couldn't get a fuel truck quickly the entire boiler would freeze up. Then we were looking at weeks to restore it to service. We would have to go in and thaw out all the pipes and there would probably be some pipe damage."

In the end, MFA Oil came through, with its truck arriving with an hour or so to spare. The coordinated relief effort between multiple organizations and state agencies was just one of many remarkable stories to come out of this disaster.

There were also fuel issues at the New Madrid Plant. Once the plant was able to restart one unit, the coal yard was still without power, and coal had to be hauled from the long-term stockpile using mobile equipment. While the supply lasted, it was closely monitored. Likewise, without power to gasoline pumps in the surrounding communities, the plant's staff took action to make sure employees could get gasoline for their cars to get to and from work.

Even with the loss of so many generating plants, the bright minds at Associated found a way to minimize the financial impact on the electric cooperatives. Knowing New Madrid would be unavailable for weeks, Associated staff moved up the timetable for a scheduled spring maintenance outage that would usually take place in April.

"Normally you are looking at something in the order of half a million dollars a day in lost opportunity for each one of those two units," Highley said. "So instead of that being a complete financial hit against us for that whole period of time, we just went ahead and started our spring maintenance outage. We avoided four weeks of later outage."

Highley saw a silver lining in the storm clouds that brought about the outage. "In my career of twenty-five years I have never witnessed that big of a loss on the grid capability. And yet I am just real pleased with the way everything performed. We lost obviously the ability to deliver to members who had their distribution lines down. But everybody that could take power was getting power. And that's

Repair trucks cluster under the network of towers in front of Associated Electric's New Madrid Power Plant. (Photo by Kevin Hopper, Associated Electric Cooperative)

One of the transmission towers leading out of the New Madrid Power Plant wears a blanket of ice. (Photo by Jim McCarty)

good news. Anybody that had a line still up where we could get to the distribution system, we were able to serve them.

"In the end, I think it all worked pretty well. It's kind of a preview of what we might see if we ever had a massive earthquake down there. So it was good for us to go through this and spend some time with the lessons learned."

During the repair effort, the New Madrid Power Plant was one of the few places in southeast Missouri with electricity. As the restoration effort began pouring massive numbers of workers into the area, the plant was pressed into service to house workers.

Plant employees donated cold weather gear to line workers from southern states unfamiliar with Missouri winters. They also contributed several hundred pairs of gloves as those brought by the crews wore out.

"They were really eager to show they could be good neighbors and they really wanted to welcome those guys and host them well," Highley says. "We're just grateful to be able to help. The linemen are the ones who really do the job. The power plant, we were doing our normal job. When a plant trips, we put it back on line. That's not anything unusual for us. The guys who really had the challenge are the linemen and our hats are off to them."

A new H structure is assembled on the ground and will later be hoisted into place. (Photo by Kevin Hopper, Associated Electric Cooperative)

Much of the conductor stretched or broke under the heavy burden of ice and had to be replaced. The conductor consists of a steel core wrapped with aluminum strands. (Photo by Kevin Hopper, Associated Electric Cooperative)

Tracked vehicles hauling extremely long poles were a common site in southeast Missouri as repairs were made to transmission lines. (Photo by Kevin Hopper, Associated Electric Cooperative)

The many ditches that drain the Bootheel provided obstacles that made transmission repairs difficult. (Photo by Kevin Hopper, Associated Electric Cooperative)

The New Madrid Power Plant owned by Associated Electric and its member systems was one of three power plants knocked off line due to a lack of transmission. (Photo by Kevin Hopper, Associated Electric Cooperative)

Piles of broken insulators had to be replaced and hauled away. (Photo by Kevin Hopper, Associated Electric Cooperative)

CHAPTER 4

M&A ELECTRIC POWER COOPERATIVE, POPLAR BLUFF

M&A Manager John Farris thought the tornado two years ago that took out thirteen steel towers on a high voltage line in southeast Missouri was the worst natural disaster his system would face. "I learned that it is a whole lot worse losing wooden poles than steel towers," Farris said in the wake of the January 2009 ice storm.

Not long after Farris took over as manager of M&A Electric Power Cooperative, a crop duster flew into one of the transmission cooperative's lines. The line stretched, and the plane dropped to the ground gently enough that the pilot was able to walk away from the accident.

"That heavy plane hit the line and only one pole broke," Farris recalls. "To think what it would take to break thousands of poles I can't even imagine."

Bootheel farmers helped speed up the restoration effort by towing the heavy transmission trucks through muddy fields. This contractor crew was working on a line across southern Pemiscot County. (Photo by Jim McCarty)

So many lines were down for M&A, which supplies power to four southeast Missouri electric cooperatives, that dispatchers ran out of the yellow magnets used to show open breakers on its system map. The transmission cooperative lost thirty-nine of its seventy-five substations and was faced with replacing a whopping 2,400 poles. Damages passed $60 million and could climb higher.

Nearly ninety miles of line had to be rebuilt completely, and none of it would come easy. Those lines crossed flooded rice and cotton fields that were a sea of mud. Only tracked vehicles could move across the fields, so trucks had to be dragged from pole to pole to make repairs.

Once poles started breaking, they fell like dominoes in a line extending from the New Madrid Power Plant to Campbell in the west and Essex to the north. The only thing that stood in their way was Crowley's Ridge, a natural speed bump in the otherwise flat Bootheel. As the falling poles hit the elevation change, the resistance was enough to stop the cascading.

The H structures that hold up M&A's high voltage lines looked like another crop duster had flown down the middle, this time taking out crossarms, braces, and the tops of poles. As far as the eye could see in the flat land of southeast Missouri, mile after mile of damaged poles stretched to the horizon. Heavy transmission conductor lay on the ground and across roads. Piles of mangled hardware and broken insulators and an eerie fog that settled on the landscape made it look like a major battle had just been fought here.

M&A Manager John Farris shows U.S. Representative Jo Ann Emerson a chunk of ice taken from one of the co-op's downed transmission lines. Farris estimated the weight of the ice on each span at 10,000 pounds, the equivalent of hanging two SUVs from each crossarm. (Photo by Jim McCarty)

No one had ever seen anything like the damage caused to M&A's transmission lines. Normally in an ice storm, distribution systems have all the damage. A transmission cooperative might lose a few poles or some hardware. But repairs generally take a short time and backfeeds take care of any major transmission outage.

"This kind of damage was pretty much unheard of," Farris says. "We've never had anything close to it. It is unusual for a transmission cooperative to go twenty-four hours with an outage. Two days is the longest. If you go three days you are in real trouble."

Because they carry much higher voltage and larger wires than distribution systems, transmission lines are built to much higher standards. Some transmission poles are three feet in diameter at the base. Yet these were snapped like matchsticks by the heavy weight of ice.

Farris compared the weight of the ice on the lines to hanging two large SUVs on each crossarm. M&A employees calculated the weight of ice on each span at 10,000 pounds for a two-inch buildup of ice. However, some lines took on more than two inches.

"We had a lot of double bundle conductors (two wires close together) on two 345 kilovolt lines. The ice was thick enough that it bridged the two conductors. Once they were bridged the ice was the size of a volleyball," Farris said.

Work gets underway by an M&A crew on a transmission line in front of Pemiscot-Dunklin Electric Cooperative's office as ice starts to melt from trees. (Photo by Jim McCarty)

The massive damage, combined with damage to lines owned by Ameren UE and southeast Missouri municipal utilities, plunged the entire Bootheel into darkness. Pretty much everything south of Highway 60 was affected.

The first sign of trouble came when the Coast Guard reported that a line tying the Associated system to the Tennessee Valley Authority was in the Mississippi River. "At that point it didn't seem like a really big deal," said Farris, who was attending a meeting of the Associated Electric Cooperative board of directors in Springfield when the storm hit. "It's all right to lose a line. There's a lot of redundancy built into the system. But by midnight we had lost the entire system. The only lights you could see were the flashing lights of the repair trucks."

M&A Manager John Farris calls for reinforcements from sister transmission cooperatives during the early days of the ice storm. (Photo by Jim McCarty)

As the situation at M&A became apparent, the five sister transmission cooperatives in Missouri all came to the rescue. Help was on the way in less than eight hours. All five systems stripped their offices down to just skeleton crews and opened up their supply inventories in an all-out effort to restore power at M&A. That assistance, combined with contractor crews, turned what should have been a year's worth of work into a four-week blitzkrieg.

One by one, the transmission crews re-energized substations until the final one, the Wilkerson sub located behind Pemiscot-Dunklin Electric Cooperative's office, came back on line four weeks later.

Farris says it will be a year before M&A's lines are one hundred percent back to normal. Crews first will have to assess whether stretched conductor needs to be replaced. When the work is completed, the system will be better prepared for another emergency.

Fallen insulators and the conductor they held sit in an icy field. (Photo by Jim McCarty)

"When we built back we used a heavier class of pole," Farris said. "We also built storm structures every five to seven miles. This should prevent cascading." The cooperative will also keep a different inventory of material on hand.

"We were accustomed to localized problems," Farris said. "This told us we could have a much broader problem than our imaginations could ever foresee. This is probably similar to what an earthquake could do."

Within eight hours of the first sign of trouble, help from the five sister transmission cooperatives was arriving in Poplar Bluff to assist M&A crews. This truck from N.W. Electric Power Cooperative came all the way from Cameron, Missouri, north of Kansas City. (Photo by Jim McCarty)

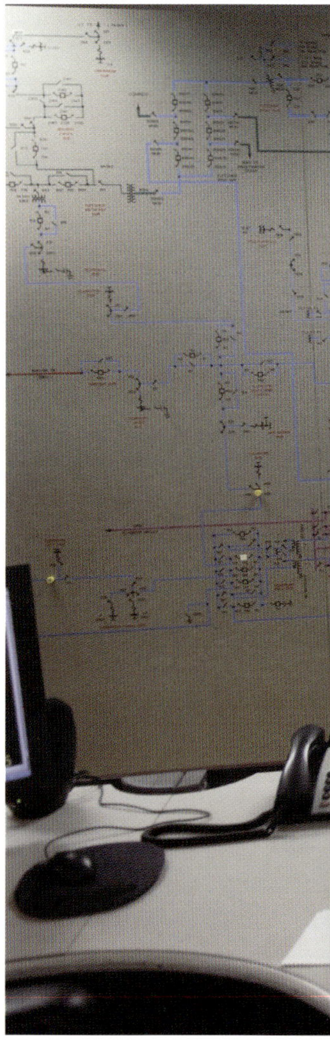

Ice covers a sign showing this to be an M&A line in southeast Missouri. (Photo by Jim McCarty)

New poles stretch across a field between Kennett and Hayti as work proceeds on replacing damaged transmission lines. (Photo by Jim McCarty)

Contractors working for M&A Power had to deal with deep, gooey mud where the lines crossed flooded rice fields. (Photo by Jim McCarty)

So many breakers were open on the M&A system that dispatchers ran out of the yellow magnets used to show their status. (Photo by Jim McCarty)

M&A employees surveying damage to the transmission grid were greeted by this sight: nearly ninety miles of shattered structures. (Photo by Jim McCarty)

Only tracked equipment could move across the muddy fields of southeast Missouri. (Photo by Jim McCarty)

A rare site was this H structure near Bragg City that stayed on its feet. Most of the poles snapped or toppled to the ground. (Photo by Jim McCarty)

CHAPTER 5
WHITE RIVER VALLEY ELECTRIC COOPERATIVE, BRANSON

The southeast corner of White River Valley's service territory was hit hardest by the ice storm. "It was as though a line was drawn down the Missouri/Arkansas line," said White River Valley CEO Chris Hamon. "Crossing Highway 5 near Gainesville was like entering another world. The landscape was beautiful, but the ice was wreaking havoc on our system. Even the most well-maintained lines couldn't withstand ice of that magnitude."

At the peak of the storm, White River Valley had around 7,400 members without power. The first of the five distribution cooperatives to restore power, White River saw power fully restored in six days.

"Number-wise, it's not the biggest outage we have had," said Jeff Pardeck, manager of member services. "But I don't recall seeing that much ice in the nineteen-plus years I've worked here."

"This was some of the worst ice I have seen," Hamon added. "What made this outage particularly difficult is that we couldn't just put the line back up. We had to start rebuilding the lines from the beginning."

With plenty of advance warning of the storm from the National Weather Service, the cooperative had all of its employees on standby waiting to see what happened. "We were prepared, waiting for it to happen, but we thought it would happen on Monday," Pardeck said. "Crews were called in and ready to go. We also alerted the media of the possibility of extended outages."

Sleet and ice fell, but the lines initially stood the strain. The next day, there were only scattered outages across the system. Then came Round Two. The first reports of outages came in around 11 a.m. on Tuesday, January 27, and steadily built throughout the day.

Some of the problems were caused by the sheer weight of the ice. But the worst damage came from tall trees, located well out of the right of way, that shattered under the weight of the ice and took power lines with them.

Dressed in fluorescent yellow-green storm gear, White River Valley crews, aided by fifty-five contractors from Southern Construction, made great progress until Thursday, when ice started melting from cedar trees near power lines. As the ice fell off, trees snapped back into lines, and the lines themselves, free from the heavy weight of ice, began rebounding. When this happened the lines touched and burned in half, or damaged equipment failed.

"Our crews fought hard against the ice and freezing conditions. Before the temperature started rising we were down to around 1,000 members without power," Hamon said. "The next day we

A big part of White River Valley Electric Cooperative's recovery efforts focused on cutting tall trees that fell into the right of way. Here, Rocky Wyman clears limbs from a line near Reuter. (Photo by Diana Bradford, White River Valley Electric Cooperative)

contacted the media to let them know that things would get much worse before power was fully restored. We knew that once it started melting, a lot of what we worked so hard to get back on would go back out. Just as we predicted, our outages spiked, going from 1,000 to nearly 8,000. Even though we expected to see our outages go up, it felt as though we were spinning our wheels."

"There was a two-hour period there where the numbers never changed," Pardeck recalled. "They would get a line back up, move to another location, and then thirty minutes later the exact same line would be back down."

Outage numbers during this time period stayed at 4,000. Any gains the crews made were undone by new outages that popped up. "If you were just looking at the numbers it appeared that we weren't making any progress," Pardeck said.

Ultimately, the hard work gained the upper hand, temperatures moderated, and the power was restored.

Hamon said that new automated meters recently installed on the cooperative's lines are proving their worth during major outages. By sending a signal from the office down through the lines, co-op

In the early hours of the ice storm roads were a sheet of ice that made travel hazardous. (Photo by Diana Bradford, White River Valley Electric Cooperative)

employees can see if a meter has power flowing through it. Even before a member calls in to report an outage, it's likely the staff already knows about the problem.

"Through the implementation of AMR technology, White River can efficiently operate during an outage even if the telephone system goes down," he said. "By remotely communicating with the meter, we can tell which side of the meter the breaker is out on. When a member calls in we can provide them with more specific information about their situation."

Hamon had nothing but praise for those who worked to get the damage repaired. "I am confident that our line crews are among the best in the business. The pride they take in their job is inspiring. They will not rest until the last member is back on."

Looking back, the cooperative employees saw many good deeds from grateful members. "The thing that sticks out the most was how many members would bring the linemen food or just try to make the situation more pleasant for them," Pardeck said.

One restaurant in Theodosia, the Sub Shack, was about to close when a group of White River Valley linemen came in looking for a late meal. The crew hadn't eaten a hot meal all day. The owner reopened and made sure the hungry linemen got fed. Elsewhere, members took their own chainsaws and cleared fallen trees so that crews had a path to the outage areas.

One of White River Valley's crews, including Larry Maric, Stan Barker, Gragg Alms, Mike Jones, Rick Burkhart, Jeff Wasson, and Bob Matthews, gets their orders in the field. (Photo by Diana Bradford, White River Valley Electric Cooperative)

Working high above the ground from a bucket truck while the surrounding woods is cloaked in ice redefines the word *cold* for White River Valley lineman Rocky Wyman. (Photo by Diana Bradford, White River Valley Electric Cooperative)

Thank-you notes from grateful members flooded the office in the days after the storm damage was repaired. "You guys rock!" wrote one member from Hollister. "Our power went out around 4:30 p.m., and a crew was here within the hour ready to fix the problem. I know y'all are prepared to work in extreme temperatures and such, I just wanted you to know it didn't go unnoticed! Our power was back on by 9:45 p.m. and other limbs that could've led to additional problems were eliminated. The fireplace was lit, as well as several candles, so we weren't suffering, but it was greatly appreciated that you were here to help on such a bitterly cold night."

The storm did about $1.2 million in damage to White River Valley's system.

'allen ice-covered limbs frame a
Vhite River Valley crew working
n the woods. (Photo by Diana Bradford,
Vhite River Valley Electric Cooperative)

White River Valley's Mike Jones uses a hot stick to knock ice off power lines. (Photo by Diana Bradford, White River Valley Electric Cooperative)

Snow and ice still fall but White River Valley crews are already at work gauging the extent of the outage. This downed pole was located near Ridgedale. (Photo by Stan Barker, White River Valley Electric Cooperative)

Rick Burkhart, Mike Jones, and Jeff Wasson team up to hoist wire to the top of the pole. (Photo by Diana Bradford, White River Valley Electric Cooperative)

A crew from Southern Construction works to repair a line near Ridgedale. (Photo by Diana Bradford, White River Valley Electric Cooperative)

CHAPTER 6
HOWELL-OREGON ELECTRIC COOPERATIVE, WEST PLAINS

Employees of Howell-Oregon thought the cooperative would dodge the bullet on this storm. Early on, only sleet fell across the system, prompting around three hundred members to lose power, nothing the cooperative's own crews couldn't handle. But just a degree or two temperature swing can make all the difference in a Missouri winter.

"I think it was around three o'clock on Tuesday, the temperature warmed up to twenty-nine degrees and the sleet changed to freezing rain," says Dan Singletary, manager. "We stood in the dispatch room and for the next two-and-a-half hours, we watched feeders just start clicking off on our system. All we needed was a few more hours and we would have come out of this OK."

Instead, heavy ice blanketed the system, collecting on tall trees and power lines alike. At the height of the storm, around 11,500 members were without power. The damage got worse around the Arkansas border.

Howell-Oregon had a forest of broken poles like this one in the wake of the ice storm. (Photo courtesy of Howell-Oregon Electric Cooperative)

"We had two things happening," Singletary says. "First, we had ice loading. When it got to an inch or bigger on our lines, that's when things started going down. The system did really well up until that point.

"The other thing, we had trees fall that were located out of our right of way. Just the weight of ice on them caused them to come crashing down on the power lines. Pines, oaks, everything. It was ugly."

Like all of Missouri's electric cooperatives, Howell-Oregon has an emergency plan that has been finely honed by years of experience. In the plan, duties are assigned. Contingencies like where to house crews coming in to help and how they will be fed are covered, as is the "Procedure for Emergency Assistance" based at the statewide office in Jefferson City.

With an early warning from the National Weather Service that this storm could be a bad one, Howell-Oregon got a head start on the repair effort.

"I got my linemen in and we started making plans before it hit," Singletary said. "I told them we are planning for twenty poles to go down and we are planning for 2,000 poles to go down. So we were arranging for crews and material before the storm even hit us."

Howell-Oregon Manager Dan Singletary gives workers a pep talk while they eat breakfast. (Photo courtesy of Howell-Oregon Electric Cooperative)

Every employee knew their role in the restoration process. Even spouses were brought into the effort, with the manager's wife, Rhonda, cooking supper for hundreds of linemen on several occasions. Churches, civic clubs, the local Bikers United chapter, the Kiwanis, and others also pitched in to feed the army of workers.

"We had all sorts of volunteers," Singletary says. "It was a true community effort."

Singletary says the cooperative's plan was successful. "We've never had to use it to this level but it worked and it worked well. We had every employee involved in this somehow."

The last few hundred members were the hardest to get back on line. Howell-Oregon crews and their helpers had to reset 450 poles in the rocky soil. Sometimes crews would set three or four poles just to restore power to one member's home, then drive many miles to the next outage.

If not for its devastation, the thick blanket of ice on everything would have been a thing of beauty. (Photo by Diana Bradford, White River Valley Electric Cooperative)

Howell-Oregon's trucks stand ready for action. Crews headed out long before the sun came up. (Photo courtesy of Howell-Oregon Electric Cooperative)

Unlike the crews working in the flat Bootheel, those helping the Howell-Oregon recovery effort had to contend with steep hills and rugged terrain. As the ice melted off, it turned from workers slipping on the ice to workers getting stuck in the mud.

Electric co-op employees learn something new each time a major outage takes place. This time, contractor crews who came from Texas brought with them some equipment that proved extremely useful in the Ozarks. The contractors had skid steer loaders with tracked undercarriages set up for line work. One machine drilled holes, while the other had a manipulator arm used to set poles.

Using these, workers could saw through fallen timber and use the pole setter to move the sections of logs out of the way. In this manner the workers literally sawed their way into the right of way in places where a truck would be hard pressed to follow.

With crews working late into the night on Saturday, February 7, the cooperative managed to reenergize all of its members the next day.

In the days after the repairs were completed, thank-you letters and cards poured in from members. One member from Caulfield wrote: "While we were bemoaning our personal situation of candle lighting and heating bath water on a propane fish cooker, and having to charge our cell phones and laptop in our car (in a warm, dry garage), we never forgot that a bunch of guys were out there in cherry pickers in the sleet and freezing rain doing what they agreed to do when they hired on. Given the conditions and the enormity of the task, the results were little less than miraculous. In an age of over-pay for under performance, we want to say 'God bless the thinkers and doers at Howell-Oregon for an outstanding job.'"

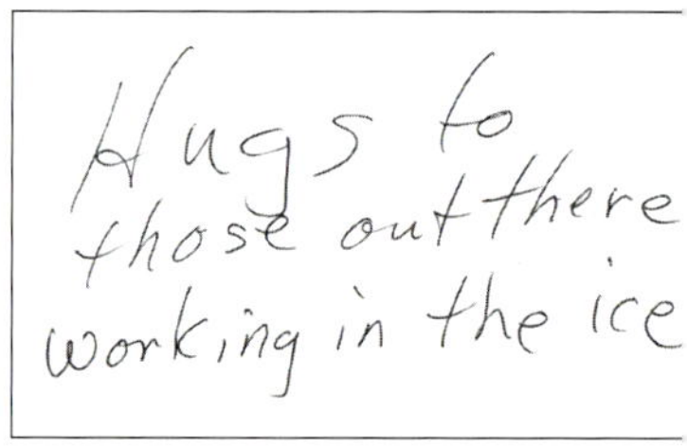
Hugs to
those out there
working in the ice

One of many letters from members thanking Howell-Oregon workers for their hard work

Many homeowners came home to find broken poles in their yard and no power in their homes. (Photo courtesy of Howell-Oregon Electric Cooperative)

Wind left this strange pattern on the ice that collected on lines two to three inches thick. (Photo by Jim McCarty)

Sometimes the poles got revenge by falling on the trees, though usually it was the other way around. (Photo courtesy of Howell-Oregon Electric Cooperative)

The crossarm on this pole hangs at a crazy angle after the weight of the ice broke the bolt off its brace. Fixing minor problems like these consumed a lot of time for the cooperative's linemen. (Photo courtesy of Howell-Oregon Electric Cooperative)

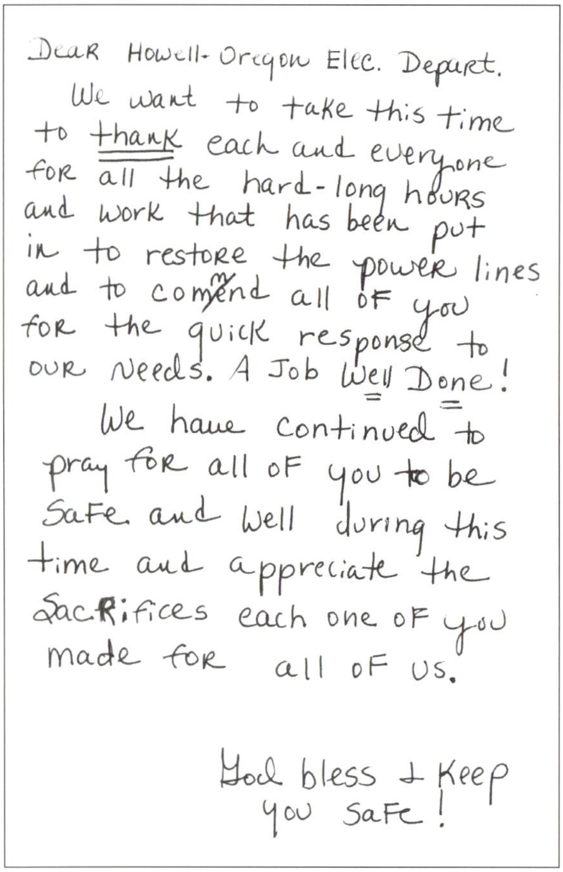
Dear Howell-Oregon Elec. Depart.

We want to take this time to thank each and everyone for all the hard-long hours and work that has been put in to restore the power lines and to commend all of you for the quick response to our needs. A Job Well Done!

We have continued to pray for all of you to be safe and well during this time and appreciate the sacrifices each one of you made for all of us.

God bless & keep you safe!

Where bucket trucks couldn't reach, linemen put on their hooks and climbed icy poles like this one. (Photo by Jim McCarty)

Right of way cleared for normal conditions fell victim to tall trees that broke and fell into power lines, taking hundreds of poles with them. (Photo courtesy of Howell-Oregon Electric Cooperative)

Howell-Oregon Manager of Operations Jack Hawkins gives out assignments to crews during their morning meeting in the cooperative's boardroom. With so many outside crews assisting local linemen, close coordination was vital to safe work. (Photo courtesy of Howell-Oregon Electric Cooperative)

Hungry cattle hope the linemen working in this field leave them some hay. This photo was shot before melting ice turned fields into a sea of mud. (Photo courtesy of Howell-Oregon Electric Cooperative)

Any home with trees in the yard needed a chainsaw after the ice storm rolled through. (Photo courtesy of Howell-Oregon Electric Cooperative)

CHAPTER 7
SEMO ELECTRIC COOPERATIVE, SIKESTON

With so many miles of line on the ground, SEMO Electric Cooperative's manager of communications Glen Cantrell likened the recovery effort to building a new power line from St. Louis to Memphis—in three weeks. SEMO's problems were trees in the north that snapped off and fell into power lines and heavy ice combined with loose, sandy soil in the south that allowed poles to topple in waves.

To repair the 3,800 poles that were damaged, the cooperative enlisted the help of more than nine hundred additional linemen from electric cooperatives and contractors. Besides the line workers who typically come to the aid of a stricken electric cooperative, SEMO also borrowed warehouse workers to keep track of the massive amounts of materials that arrived by the truckload, and staking engineers to mark the path for the new lines.

"We went from sixteen linemen and a week later we were running 940 workers," said Reuben Jeane, SEMO Electric's manager. "For three days there was no electricity anywhere. It was terrifying, but we survived it."

Ice clings to every surface of this SEMO Electric Cooperative pole. (Photo by Glen Cantrell, SEMO Electric Cooperative)

Across the cooperative's service area, hardly a pole could be seen standing. Along Highway 60, for example, from Idalia to Sikeston, there were twenty-five snapped transmission poles before a single pole could be found standing. As the poles fell into thick slushy ice, they disappeared from view. Only jagged stumps sticking from the ground showed where a power line once stood.

With power off in Sikeston, gas was in short supply. Employees shared the same fate as members, with few having power at their homes.

Cantrell says his home in Sikeston was without power for twelve days. "I live in a neighborhood that has a lot of old trees," he said. "I remember the first few nights, the trees were coming down. It was like a shotgun blast every minute. Crack! And then crash. I can't imagine what those big poles sounded like coming down."

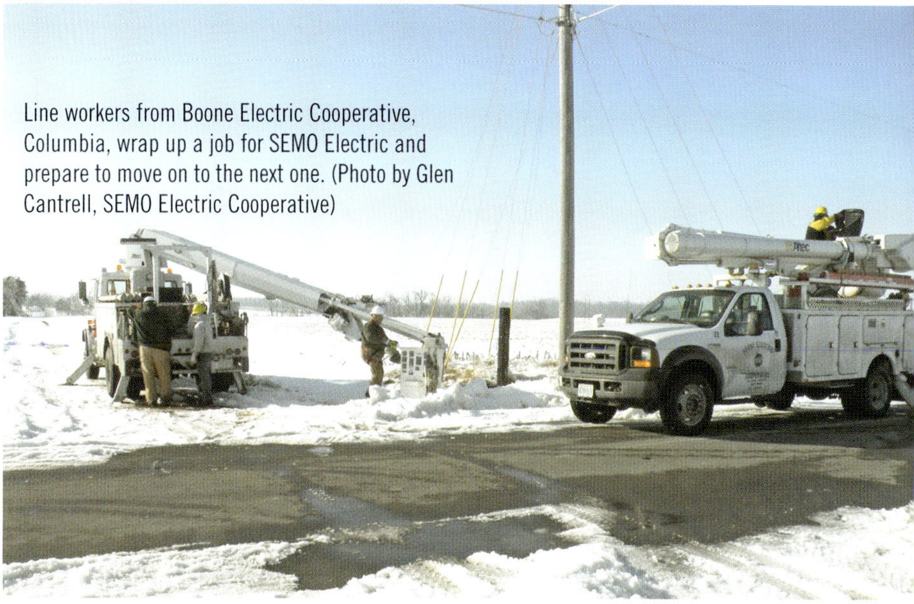

Line workers from Boone Electric Cooperative, Columbia, wrap up a job for SEMO Electric and prepare to move on to the next one. (Photo by Glen Cantrell, SEMO Electric Cooperative)

Normally during a major outage situation, linemen work from the substation out. As they repair damaged feeder lines, large numbers of meters start turning again. But with most transmission lines feeding southeast Missouri also damaged, crews could not see their labors bearing fruit. It would be many days before transmission could be restored to substations.

The cooperative seemed to have turned a corner on February 10 when it saw outages dip from a high of 11,000 to below 700. Then powerful winds blew into the area and undid much of the work.

With gusts of up to sixty miles per hour pounding the service area, SEMO's staff watched outage totals bounce all over the place. Crews would energize a line, only to see another go down in a day of tail chasing.

The soft, sandy soil of the area liquefied under the double burden of wind and rain. Poles that had been recently replaced once again toppled to the ground. In some areas, poles that earlier had withstood the burden of ice toppled into the mud.

Jerry Dockins, a twenty-four-year veteran of the cooperative, was just amazed at the level of destruction caused by the heavy ice. "I thought the ice storm of 2008 was the worst one I would ever see. 2008 was a piece of cake compared to how this one is going to turn out to be."

The area southeast of Dexter was especially hard hit by the storm. Virtually no poles were left standing in the area.

Dockins spent much time traveling the system delivering lunch and materials to the crews in the field. He expressed awe at the devastation he witnessed. "I've seen miles and miles of wire and poles lying on the ground. Transmission poles lying on the ground. 345 kv transmission lines lying on the ground. H fixtures down just as far as you can see in both directions.

"I talked to an old man who is ninety-one years old and he was born and raised at Kewanee. He said he has never in his life seen anything as bad as this ice storm. He would know."

With so many extra workers helping out, the logistics of feeding crews taxed the cooperative's emergency preparedness plans. SEMO directors came into the office daily to prepare thousands of sandwiches, many bringing their families with them. "They were not only fixing lunches, they were delivering them too," Cantrell said. "We were packing around a hundred lunches each and driving around to different locations to hand them out to crews."

Area churches volunteered for whatever job they could do. The Miner Baptist Church, which is served by SEMO Electric, took in laundry daily from the workers. They divided it up between themselves and other area churches, washed it, and returned the clean clothes the same day. "These guys in from another town, they would have to go find a laundromat," Cantrell said. "But when would they go? These Pike guys (contractors) told us they had never had anyone do that for them before."

Employees of Pike Electric, a contractor called in to assist at SEMO Electric, drag heavy wire into place for restringing. (Photo by Glen Cantrell, SEMO Electric Cooperative)

Communities came together to help total strangers eat and stay warm. Everywhere, the disaster brought out the best in people.

Despite the crisis, some things could not be put off. That was the case with one lineman who worked for Pike Construction, one of the largest electrical contractors in the nation. Pike brought in crews from all across the country, including one worker who was supposed to get married the week he was sent into the fray. Rather than put off the wedding, his fiancée made the trip to Sikeston. The couple found a local church and a preacher and tied the knot in Missouri. One can only hope they later had a nice honeymoon somewhere warm.

In the midst of the ice storm, SEMO manager Reuben Jeane plopped in a chair in Jerry Dockin's office, weary from two weeks of intense activity and little sleep. From a box on the floor he pulled out a hat marked "SEMO Ice Storm 2008."

"Next time we need to leave off the date," he remarked. The 2009 ice storm marked the third major storm that hit the ill-fated southeast Missouri co-op. The first was an ice storm that hit around Valentine's Day in 2008. Then in September of that year, the remains of Hurricane Ike wreaked havoc on the cooperative's lines. When the third storm hit in January 2009, the cooperative's crews were battle-tested and itching for a fight—with Mother Nature.

On February 16, they managed to restore power to every member capable of taking it. Work would continue on agricultural accounts and cleanup efforts for many more weeks.

This closeup of a pole top shows the heavy ice that stuck to everything. (Photo by Glen Cantrell, SEMO Electric Cooperative)

While the Missouri Highway Patrol stops traffic, crews helping SEMO Electric get back on its feet drag conductor across Interstate 55. Many such crossings had to be made before the effort was complete. This one took just eighteen minutes. (Photo by Glen Cantrell, SEMO Electric Cooperative)

Boone Electric lead lineman Jamie Conrow tells about his experience during the ice storm while reporters from KOMU-TV tape him. (Photo by Glen Cantrell, SEMO Electric Cooperative)

Before SEMO Electric's employees could repair downed lines, they first had to de-ice trucks and other equipment. (Photo by Glen Cantrell, SEMO Electric Cooperative)

Volunteers prepared thousands of lunches and helped deliver them to the linemen in the field. Pictured here are (from left) Debbie Deneke (wife of SEMO board member James Deneke), Beverly Jeane (wife of manager Reuben Jeane), and Julie Deneke (daughter of board member James Deneke). (Photo by Glen Cantrell, SEMO Electric Cooperative)

SEMO workers had to deal with broken poles and ditches full of water. (Photo by Glen Cantrell, SEMO Electric Cooperative)

New poles arrived around the clock as SEMO and other electric cooperatives replaced thousands that were destroyed by the heavy ice. (Photo by Glen Cantrell, SEMO Electric Cooperative)

SEMO Manager Reuben Jeane gives U.S. Representative Jo Ann Emerson an update on the extent of the damage to his cooperative during a visit to Sikeston made by the southeast Missouri legislator. Congresswoman Emerson provided invaluable assistance to cooperatives in her district during and after the ice storm. (Photo by Jim McCarty)

Farmers with their huge tractors saved the day for SEMO Electric by dragging trucks across muddy fields. (Photo by Glen Cantrell, SEMO Electric Cooperative)

CHAPTER 8

OZARK BORDER ELECTRIC COOPERATIVE, POPLAR BLUFF

"I've never seen anything like this," says Ozark Border Manager Stanley Estes, who has worked for the cooperative for thirty-two years. "It is by far the worst. We've got 170 of these Pike linemen who work a lot of hurricanes. The co-op linemen, they say the same thing. This damage is as bad as you see in a hurricane if not worse."

For emphasis, Estes points to the 3,000 poles the cooperative lost. That compares to 1,000 replaced by Dixie Electric Cooperative, a Louisiana system helped by Ozark Border crews after Hurricane Katrina.

From Williamsville in the north to the Arkansas Border near Doniphan, Ozark Border saw major outages. At the peak of the storm, more than 28,000 were without power. That number went down in a hurry as the repair effort gained momentum.

This sign thanking Ozark Border employees for their hard work during the storm showed up anonymously overnight at the headquarters in Poplar Bluff. (Photo courtesy of Ozark Border Electric Cooperative)

Employees, most without power at home themselves, worked around the clock for the first few days in order to get ahead of the outages. Chuck Reinbott, for example, worked for three days before he got a chance to go home to tend to his livestock. His wife had bought a generator but he had no chance to hook it up for her. He brushed his teeth without water. When he crawled into bed, exhausted, it was like "slipping between two sheets of ice," he said. He brought his electric razor to the office to shave with but never found the chance.

Keith Stoelting, the co-op's eastern area supervisor, worked twenty-five hours before he got any rest. A call from Estes—"Chief, here it is"—woke him up at 2 a.m. on January 27. Driving the gravel roads east of Poplar Bluff, Stoelting pointed out areas where his crews had just built new lines or moved old ones away from heavily forested areas. The heavy ice knocked down new poles and old, apparently choosing its victims randomly.

As night fell in Stoddard County, the darkness was complete save for a puny light poking up into the black sky. The lonely light turned out to be the spotlight from a Barton County Electric Cooperative truck. Its crew, unwilling to call it quits until their assigned stretch of line was rebuilt, worked until after 9 p.m. before heading back to Poplar Bluff for some supper. Their efforts were made more difficult by the swampy soil. When they drilled holes to set poles into, water quickly seeped in and often made the holes collapse.

Kenny Morris, a lineman for Barton County Electric Cooperative, gets ready to put safety grounds on a damaged section of Ozark Border's line near Qulin. (Photo by Jim McCarty)

Back at the office, outage calls poured in. Branch offices were closed and the employees were brought in to the Poplar Bluff headquarters to handle the volume. One room was devoted to a bank of phones, and office staff worked in shifts to handle the thousands of outage calls. Only in the wee hours of the morning did the calls let up, giving employees time to get organized for the next day's work.

Teresa Fansler, the cooperative's executive secretary, had the duty of finding motel rooms for incoming crews. "I've never rented an entire hotel before," she was heard to remark. With so many people without electricity and so many crews coming in to repair the damage, some linemen had to drive as far away as Van Buren to find lodging.

One of the more interesting experiences from the storm came from a crew on loan from New-Mac Electric Cooperative working in Ozark Border's service area. Grateful Ozark Border members took it upon themselves to feed the crew members, who were no strangers to ice storms back home. "Who wants to be the first to be kissed?" asked an elderly lady, no doubt ecstatic to see the repairs coming so close to her home.

On February 11, outages were under 1,000 and Estes could see the proverbial light at the end of the tunnel. "We were just about to get our hands around it," Estes says.

Then powerful winds tore through the cooperative's service area. Poles that had been previously replaced leaned or fell down in the waterlogged soil southeast of Poplar Bluff. Three substations and five feeder lines went down. Tractors and other heavy equipment were tied to poles to keep them standing.

While the new problems were quickly repaired, crews felt like they were taking one step forward and two steps back. The setback added at least a day to the repair efforts, but Ozark Border could

Eastern Area Supervisor Keith Stoelting works to clear fallen lines from a road near Campbell. (Photo by Jim McCarty)

Billing supervisor Pam Overfield organizes piles of outage reports so they are ready to hand off to crews working to restore power. (Photo by Jim McCarty)

Brian James, a lineman for Ozark Border Electric Cooperative, works to clear icy limbs from a line west of Dexter. (Photo by Jim McCarty)

Kathleen Bader with St. Teresa Catholic School, left, and Teresa Fansler, Ozark Border's executive secretary, hold a sign made by students at the school in Glennonville. (Photo courtesy of Ozark Border Electric Cooperative)

A crew from New Mac Electric Cooperative, no strangers to ice storms, stops traffic on Highway 53 in order to move a new pole into position near Qulin. (Photo by Jim McCarty)

finally say it had restored power to all on February 17.

Not long after power was restored, employees arriving at the office in the morning were greeted by the sight of a huge banner that read "Thank you to the men and women of Ozark Border for a job well done. From the Citizens of Butler County." The sign was hung anonymously after hours.

Two appreciation dinners—one in Poplar Bluff and another in Doniphan—were held to recognize those who did so much to restore power. Other signs of appreciation included a jumbo poster from St. Teresa Catholic School in Glennonville that read "Thank you for joining hands to help Southeast Missouri," complete with construction paper hands from all the students.

Of the many thank-you letters the cooperative received, one really spoke to the gratitude members expressed to the line workers, both local and from out of town. "I now know how the settlers in olden times felt when the cavalry came to their rescue. That's just how we felt when we saw the electrical trucks and workers coming down our road. Thank you for all the hard and dangerous work you did in extremely cold and harsh weather."

Barton County linemen David Engelage, Kenny Morris, and Rodney Onstott open breakers in an Ozark Border substation prior to working on a section of line in Butler County. (Photo by Jim McCarty)

A crew from Laclede Electric uses a tracked vehicle to replace an Ozark Border pole in a muddy field. (Photo by Robert Henson, Ozark Border Electric Cooperative)

Ozark Border Manager Stanley Estes, left, and Data Processing Supervisor David Kirkman survey the crew assignments on a dry-erase board in the co-op's operations center. (Photo by Jim McCarty)

Bitter cold, icy roads, flooded ditches, and muddy fields all made working conditions terrible for the linemen who worked the ice storm of 2009. (Photo by Jim McCarty)

Rows of poles, uprooted and snapped in half, lined roadways across the cooperative's service area. (Photo by Jim McCarty)

In order to handle the heavy volume of callers reporting outages, Ozark Border closed its branch offices and brought employees into the headquarters in Poplar Bluff to staff a phone bank. Calls came in around the clock. (Photo by Jim McCarty)

CHAPTER 9

PEMISCOT-DUNKLIN ELECTRIC COOPERATIVE, HAYTI

By far the worst damage done by the storm was in the two Bootheel counties served by Pemiscot-Dunklin Electric Cooperative. In the wake of the storm, virtually the entire Bootheel was without electricity.

"We had been through ice storms before, or at least we thought we had," says Tim Davis, the cooperative's operations manager. "But this one was so massive. When we looked around at the destruction, there were more poles down than there were still standing. And then when we saw the transmission down, it led to a real hopeless feeling."

Davis says progress was measured by inches in the early days of the storm. "For a while we were just clearing roads and driveways. We couldn't put anything up. When we tried at first to stand something back up the ice load was so massive we would break things. The crossarms broke off when we picked them up. That was real sad because you want to get people back on."

A lineman for Louisiana's Concordia Electric Cooperative works to right a leaning Pemiscot-Dunklin pole near the Bootheel town of Senath. (Photo by Jim McCarty)

Two weeks later, only three of the cooperative's substations had power to them due to the massive amount of damage done to the transmission lines owned by M&A Power Cooperative that serve Pemiscot-Dunklin. Service had been returned to 2,000 members, primarily in the north end of the system, but each one came at great difficulty as trucks had to be dragged from pole to pole through muddy and flooded rice fields.

When roads were cleared enough for an assessment, Manager Charles Crawford found ninety percent of the system was on the ground and hardly a meter turning. Early on, Crawford called for 1,000 additional workers to help the co-op get back on line. In short order, he got what he was wanting.

To house this army, the cooperative turned to the University of Missouri Delta Center, which readily turned over its Rone Exhibition Hall to the effort. Crews were also housed at the New Madrid Power Plant, the first time this has ever been done.

Crawford called on Shorty's, a restaurant located in Holcomb, to feed the crews. Initially, owner

Ron "Shorty" Cross offered to do this for nothing. But Crawford enlightened him to the huge appetites of electric cooperative linemen working in bitter cold and convinced him he better send a bill. For three weeks Shorty's served massive amounts of food delivered to the Rone Center, the co-op's office, and at the restaurant.

Ice hangs from the sign on the front of the Pemiscot-Dunklin Electric Cooperative office. (Photo by Jim McCarty)

No single person could coordinate the work of all those additional helpers. Enter Jerry Wellington, the operations manager for Crawford Electric Cooperative, who was loaned to the effort by an understanding manager. Wellington stayed with the effort until the final days, freeing Pemiscot-Dunklin's Davis for more pressing issues.

On Sunday, February 8, Crawford surveyed the destruction from the air. "The damage is incredible," he said after the flight. "But you can sure see a lot of action taking place. At night you can really see a lot. There are truck lights flashing all over."

Those without power included all of the cooperative's employees and the office. With power out at the office, a generator was hooked up to run lights and a few essentials. For several days there was no heat in the building except for a propane stove in the lobby. Running water gave out and portable bathrooms were set up outside. With temperatures in the single digits, trips here were understandably short. That all changed when a larger generator was installed, allowing all the office equipment to be operated.

One of the last places to be reconnected in the ensuing weeks was the co-op's office and the manager's home.

The outage at Pemiscot-Dunklin set several records for Missouri electric cooperatives. Broken poles totaled around 7,500. And the 1,000 additional crew members from cooperatives in Missouri, Iowa, Mississippi, and Louisiana, along with contractors released by Ameren UE, marks the largest mobilization of men and equipment sent to a single Missouri electric co-op.

Many of the workers coming from southern states suffered in the bitter single-digit cold. Hearing of their plight, Pollack Broadcasting radio personalities Steve Tyler and Monte Lyons put out an appeal for ski masks. Listeners responded by the hundreds, flooding locations in Kennett and Caruthersville with donations. The two devoted their programming, heard throughout the Bootheel, to helping residents cope with the power outages for the duration of the storm.

Pemiscot-Dunklin Manager Charles Crawford, left, gives Senator Kit Bond an update on his cooperative's recovery efforts during the senator's trip to Kennett. (Photo by Jim McCarty)

Ultimately, the efforts of so many workers paid off. The crews helping Pemiscot-Dunklin averaged a mile of line per crew per day. One Louisiana electric cooperative crew leader apologized to Crawford as he ate supper for what he thought was a poor day's effort from his crew. He told the manager they had gotten a late start, and only set forty-six poles. He promised to do better the next day.

As M&A reenergized substations, the distribution lines were ready to take power. Celebrations broke out around the Bootheel as electricity began flowing to places like Hollywood, Peach Orchard, Gobler, and Braggadocio, to name just a few of the more colorful town names.

Crawford, a twenty-six-year veteran of the cooperative, says the destruction done to the Pemiscot-Dunklin system is the worst he has ever seen. "It's just unreal," he says. "Poles came uprooted. Poles that were new broke. Poles that could not hold the weight of the ice broke. Lines came down."

Most of the linemen working the outage agree with the damage assessment. Even those who worked Hurricanes Katrina and Rita in the Gulf Coast say the damage here was more extensive. "This was a hundred-year storm," Crawford says.

Jake Fisher, Pemiscot-Dunklin board president, put things in perspective: "It took seventy years to build this system," he said. "It took nature three days to destroy it. And we built it back in three weeks."

Macon Electric lineman Terry Blaise untangles fallen lines in a field near Hayward. The veteran lineman said the damage he witnessed in southeast Missouri made the 2001 Macon Electric ice storm seem like "a walk in the park." (Photo by Jim McCarty)

Concordia Electric Cooperative linemen prepare to remove a shattered pole from a cotton field near Senath. The Louisiana linemen, and their counterparts in Mississippi, were quick to respond to the call for help as a way to repay Missouri crews for their assistance in the wake of Hurricane Katrina. (Photo by Jim McCarty)

Trucks are gassed up before the sun rises behind the Pemiscot-Dunklin headquarters. (Photo by Jim McCarty)

Pemiscot-Dunklin's Charlie Crawford, left, and Tim Davis use a map of the system to show one of the many contractors where his crew will be working. The cooperative went from twenty-five employees to more than a thousand in just a few days. (Photo by Jim McCarty)

New poles go up in the Hayward area as linemen from Platte-Clay Electric reset three-phase lines for Pemiscot-Dunklin. Linemen averaged rebuilding a mile of line per crew per day. (Photo by Jim McCarty)

Thank-you notes and drawings from area school kids poured into the Pemiscot-Dunklin office and were prominently displayed. Schools also made valentines for the linemen working on February 14.

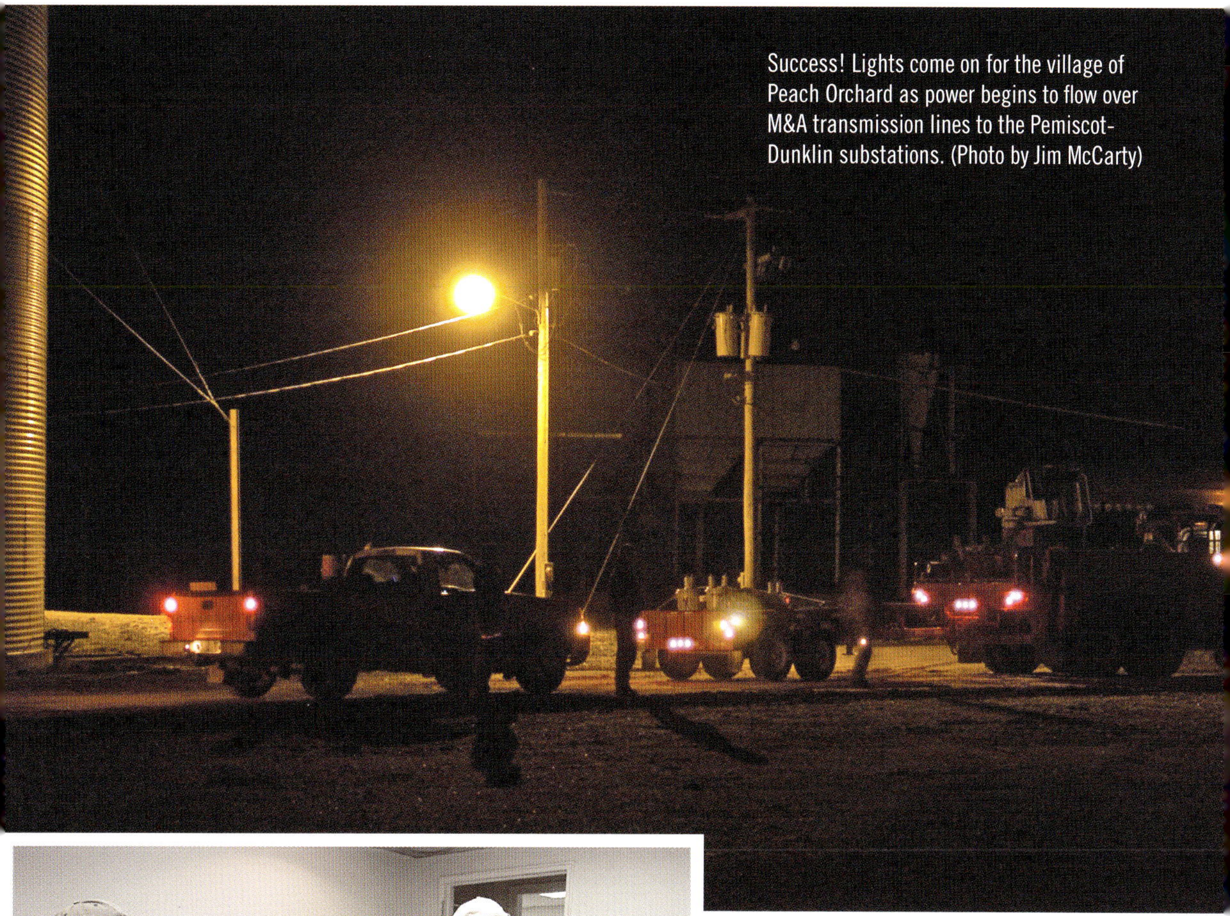

Success! Lights come on for the village of Peach Orchard as power begins to flow over M&A transmission lines to the Pemiscot-Dunklin substations. (Photo by Jim McCarty)

Jerry Wellington, center, hands out work assignments that are relayed to crews by Tim Davis, left, and Mike Markham, right. Wellington, who is Crawford Electric's operations manager, was on loan to Pemiscot-Dunklin for the duration of the ice storm. Markham coordinates sending assistance to cooperatives in need for the Association of Missouri Electric Cooperatives.

Tim Davis feels the strain after three weeks of coordinating repairs to heavily damaged lines at Pemiscot-Dunklin. The operations manager answered hundreds of calls daily on his cell phone and sometimes had two phones and a radio going at the same time. (Photo by Jim McCarty)

emen from Southwest Electric Cooperative restore power tside the University of Missouri's Rone Exhibition Hall. The ilding was used to house and feed linemen working for miscot-Dunklin. (Photo by Jim McCarty)

As temperatures warmed and the ice started to melt, mud instead of ice became the enemy. Tracked vehicles like this one owned by Barton County Electric Cooperative were all that could move across fields. This crew was working on Pemiscot-Dunklin's lines near Marston. (Photo by Jim McCarty)

ABOUT THE AUTHOR

Jim McCarty is the director of communications for the Association of Missouri Electric Cooperatives and the editor of *Rural Missouri* magazine, where he has worked for the past twenty-three years. During the ice storm of 2009, Jim threw his hard hat and sleeping bag in the back of his car and headed south, where he was more or less on permanent loan to the electric cooperatives of southeast Missouri. He is a certified rural electric communicator and the author of three books on the history of rural electric cooperatives.

(Photo by Keith Stoelting, Ozark Border Electric Cooperative)